I0625529

Growing Green –
A Beginner's Guide to Successful
Greenhouse Gardening

Cultivate Your Own Oasis of Abundance with
Essential Tips, Techniques, and Troubleshooting for
Novice Greenhouse Enthusiasts

By: Kathy Campbell

Text Copyright © Lightbulb Publishing

Legal & Disclaimer

Greenhouse for Dummies

Chapter 1

Overview

A greenhouse is a method or arrangement that helps us regulate climatic characteristics like humidity, temperature, light, etc. Greenhouses are made with different types of designs and structures, but the basic feature that all greenhouses have is that they have transparent roofs and walls in order to absorb heat as well as light from the sun. Some common materials are used in greenhouses for the purpose of transparent walls and roofs.

1. Rigid plastics that are made of polycarbonate

2. Plastic films that are made of glass panes

3. Plastic films that are made of polyethylene

How does a Greenhouse work?

The concept is that when the interior of the greenhouse is exposed to the sunlight, the temperature rises and protects the plants from the cold weather and temperature.

The purpose of greenhouses is, of course, growing plants, vegetables, flowers, fruits and also for transplanting purposes.

There are no size limitations for greenhouses; it all depends on the purpose and on the budget and availability of the space. A greenhouse can be a small shed structure or a huge facility dedicated for plants, flowers, or fruits. The large greenhouses are usually equipped with high-tech in order to maintain the large number of units and the temperature around them.

History of Greenhouses

Roman Empire

The history of greenhouses goes a long way back to the Roman Empire when the first attempt to make an artificial environment for growing plants was made. Until then, obviously, the agricultural operations were planned according to the climate and weather conditions. The first attempt was made when the Roman Emperor at that time (Tiberius) was having health problems, and the physicians suggested the emperor have one cucumber every day. Since cucumbers are not grown at the same pace throughout the year, they have their peak time; however, since the emperor needed to have one of those every day, the Romans designed and prepared an artificial environment in order to make it available for the emperor throughout the year.

Wheel carts were used to plant and put cucumbers under the sun on a daily basis, and at night, they were taken inside to keep them warm. The grown cucumbers were kept in cucumber houses ()or selenite sheets.

15th-century Korea

The next very noticeable advancement in greenhouse design was when it was designed in Korea in the 15th century at the time of the Joseon Dynasty. Sangayorok was a manuscript by Soon Ui Jeon in which he described the very first artificially heated greenhouse structure. Soon Ui Jeon was the royal family's trusted physician, and his manuscript "Sanyayorok" was an attempt to provide authentic and important knowledge for housekeeping and agriculture.

Under the agricultural techniques section in the manuscript, the physician talked about building greenhouses that were able to produce plants and vegetable crops in the winter season.

The design of Korean greenhouses had an added system known as Ondol. This is a heating system in which the temperature and heat are controlled by using a flue pipe that is connected to a heat source placed underneath the floors. Moreover, besides Ondol, for an additional source of humidity and an increase in temperature, a cauldron filled with water was heated and placed in the greenhouse. The Korean-style greenhouses also had the typical heating methods like windows and walls to capture light and heat, but better control was achieved due to the furnace. Sources of the Joseon Dynasty say that there were greenhouse-type structures using the Ondol system in 1438 for the Mandarin trees' heating and production.

17th century

The earliest greenhouses include the experiments done in the Netherlands and England in the 17th century. The early attempt happened to be quite difficult, with a struggle to maintain them in the nights and between weather changes. Providing enough heat was quite a bit of a challenge for them in those early stages. In the UK, the first greenhouse was completed in 1681 in the Chelsea Physic Garden. Today, there are many greenhouses in the Netherlands that are considered to be the largest greenhouses in the world, which are also responsible for producing a large number of vegetables on an annual basis.

Meanwhile, experiments in greenhouse design continued during the 17th century in Europe. By that time, the technology was upgraded, and better techniques and better glass materials were improved.

19th-century England

A design that was quite prominent in the 19th century was greenhouses having enough height and space for large trees. These greenhouse designs are called palm houses. These structures were built in parks and public recreational outdoor places. The palm houses actually are a noticeable progress of the 19th century in terms of developments of iron and glass architecture. Along with the "palm houses," the technology was widely used in big markets, large buildings, exhibition halls, and railway stations, and basically the places where large open area was needed.

Some of the noticeable palm houses in the early stages of this design were Palm House Kew Gardens (Royal Botanic Gardens, Kew London, 1848). One of the largest greenhouse structures to be built in the 19th century was the New York Crystal Palace, and, not to forget, the Royal Greenhouses of Laeken, Belgium.

20th century

The 20th century achieved a lot of milestones in history, including the milestones regarding greenhouse construction. The Geodesic Dome is one of the noticeable works of this century, along with the Eden Project |(Cornwall), The Climatron (Missouri Botanical Garden in St. Louis.), the Toyota Motor Manufacturing (Kentucky), and The Rodale Institute (Pennsylvania).

The Pyramid-shaped greenhouses also became a trend in the early 20th century for larger greenhouses. There are several pyramid-shaped greenhouses in Alberta at the Muttart Conservatory.

Structures of greenhouses progressed in the 1960s when polyethylene film sheets became available all across the country, and these sheets happened to be wider than the ones previously used. After this, the making of hoop houses became popular as they were being made by many companies and also individual growers. The costs were less as they required less expensive equipment because the purpose of hoop houses is usually for a temporary period, unlike greenhouses, which are built for a long-term vision.

The Gutter-connected greenhouses also became common and popular in the 1980s and 90s. These specific greenhouses are connected with two or more bays that are connected by one common wall or support posts row. Since the area was increased due to gutter and roof space, the heating inputs were ultimately reduced, and this proved to be efficient. These types of warehouses are widely used for production as well as for the commercial purpose of selling plants to consumers.

Uses and Functions

Greenhouses help to improve control over plants' growing environment. The common factors that can be controlled within the greenhouses are levels of light, temperature, humidity, and shade. Along with these factors, the application of fertilizers irrigation is also something to be done properly. Greenhouses can be beneficial in situations like when a certain piece of land is not producing quality outcomes. At that time, greenhouse concepts were very useful in overcoming the shortcomings of that piece of land and improving its growing quality.

Shade houses are used especially to provide shelter and shade in dry and hot climates.

For quick turnarounds, flowers/vegetables or fruits can be set up to be grown within the greenhouses in the early springs or late winters, and when the weather starts to warm, they can be transplanted outside. So, the seed trays can be stacked in the greenhouse with the purpose of transplanting them later outside.

An advantage of bumblebees is that they are natural pollinators along with other bees that also serve the purpose of pollination. Besides these methods, artificial methods are also now used for pollination in greenhouses.

As expected, the ways to manage the closed structure of a greenhouse are different and unique as compared to the outdoor methods of maintenance and production. A few things that need to be controlled are the same, such as pests, diseases, and water supply. Then, there are a few things that are not applied in the outdoor setting for obvious reasons, such as temperature control and humidity. For timely water supply, greenhouses are installed with drip lines or sprinklers. For temperature and humidity, specific methods for light and heat are required, especially in winter, for the production of warm-season vegetables and fruits.

Benefits

Increased Yield: One of the most noticeable benefits of greenhouse gas is that the yield increases as weather conditions are controlled.

Pest and Disease Control: One great benefit that the greenhouse concept serves is that it creates a wall between pests and plants. So, there is much less need to use pesticides, and ultimately, this saves money.

Environmental Control: The fact that we have the power to regulate the conditions actually allows us to make more efficient use of resources such as water.

Greenhouses play an important role in the areas of agriculture, botany, and horticulture by enabling us to control the plant environment for experimentation and plant growth.

Chapter 2

Planning the Structure for your Greenhouse

Types of Greenhouse Structures

At this point, we already have an idea that greenhouses can be of different types and designs, having different purposes along with serving their basic purpose. Some of the greenhouse structures are made for large-scale production, whereas some are actually well-suited for smaller or limited spaces.

Another thing is that the structures of greenhouses can be either independent or attached to an already constructed building or another greenhouse.

Gable-Style

The style structure of a greenhouse is a traditional design in which there are sloped roofs with straight walls, which makes it a triangular shape on the above part of the structure. The good thing is that light and heat have a lot of room inside to spread, and this structure can be built easily and can be considered to cover large or even small areas.

A-Frame

A-frame-style greenhouse structures are quite similar to the gable-style structures. In fact, they are easier to build, and the roofs are sloped to the ground, so there are no straight walls on their sides. As for light, the A-style structures are very good as they utilize the light at their maximum, but when it comes to space and airflow, there is limited room for it on parts near the ground because of the sloped roof all the way. Overall, it is preferred for storage rather than growing purposes.

Geodesic Dome

A geodesic dome greenhouse structure is a design that is recommended for areas having frequent severe weather conditions. A geodesic dome greenhouse has a very strong structure in the way it is built and assembled.

Gothic Arch

The Gothic Arch structure can be summed up as a combination of the A-style and gable-style greenhouse structure. Steel and Aluminum poles are used to form the walls of this structure and meet at the top, making a pointed roof similar to a cathedral.

The plus point of this structure is that it easily sheds rainwater and snow from it. It is assembled easily.

Hoop House

Hoop House greenhouse structure is made from semi-circular hoops. The ends of the hoops are attached to base plates or can also be fixed into the ground, and this way, they ultimately act as wall and roof support. So they are easy to construct.

They are the best choice structure to cover long rows and also provide a lot of room as compared to the A-frame style structure. The hoop house structure should be covered with polycarbonate films as they provide flexibility.

Ridge and Furrow Design

One more amazing example of a structural greenhouse is the A-frame similar design, that is, the ridge and furrow design. Basically, in this design, the A-frame greenhouse structures are connected in an organized row. Every row takes advantage of its roof, from which the rain and the snow flow off.

Sawtooth

Sawtooth is a unique structure that features extra ventilation from the windows on the lower sides of the roofs that can be opened when needed.

Uneven Span

When you want to build a greenhouse on land that is sloped, this is where the Uneven Span style greenhouse structure is best to be made. This structure is somewhat like a gable design with a little variation, having one side of the roof facing the south and much longer than the other side of the roof in order to maximize the intake and utilization of light.

Lean-to

The lean-to design of a greenhouse is basically a structure that is built onto an existing building or wall and is made as an arched or sloped design. It is preferred that it faces to the south as we can imagine that this design is expected to be used for urban areas and properties where there is not little space. You might think that one side of the greenhouse is dark because it is attached to the wall or a building, and this is a con, but the plus point is that the building provides stability, and the heat keeps the greenhouse warm.

Other Types of Structures

Shade Houses

A shade house is a type of structure that is covered with woven or other materials to let the sunlight in and let the air and moisture pass through the space between the gaps. Not only that, the covering material is also used for the purpose of providing an environmental adjustment like reduction in light or protection in severe weather conditions. Shade houses are mostly good for warmer regions. There is no fixed size for these structures as it depends on the crop type being maintained and produced.

Screen Houses

Rather than glass or plastic, screen houses are constructed with insect-screening material. They provide environmental adaptation, pest exclusion, and weather protection from severe weather conditions. They are widely used in warm or tropical regions to get some of the benefits of greenhouses.

Crop Top Structures

Crop-top greenhouse structures have roofs but do not have apparent walls. The roof is just held by metal or wooden rods fixed into the ground. The roof material that covers the crops can be glass, shade cloth, plastic, or an insect screening. The purpose of these structures is to protect the crops in times of rain or when there is less light.

What did we learn by knowing about the different structures of greenhouses?

Investments in greenhouses are based on technology and efficient use of materials. The better the technology and the structure (depending on the purpose), the better you can control the growing conditions. The ability to closely regulate the growing environment has a direct impact on the crop's productivity and overall health.

Chapter 3

Construct/Setup Your Greenhouse

Check your local authorities' zoning laws and regulations and inquire about the necessity of a building permit before you begin. Assemble your greenhouse safely and with the assistance of a friend. Assistance lifting and holding the parts steady will be needed in order to execute this properly. If you'd like to build a large greenhouse but lack the space to lay a foundation, a portable greenhouse can still provide your garden plants with the same advantages. Crop rotation during the seasons can also be considered, keeping in mind the greenhouse's movable design. At the end of a season, use the greenhouse to extend the growth of summer flowers and vegetables. The plants can be added to the compost pile when they are finished and when you make sure that they have no symptoms of pests and diseases.

Let's start,

1. Selection of a Location

Select a location that gets plenty of sunlight throughout the day and has good drainage. The greenhouse should be positioned to face south and an open sky. By doing this, you'll get as much sun as possible during the winter. Additionally, the location must be level or nearly level.

Most gardeners also select a location that is close to a water and electricity source.

2. Foundation

Although not all greenhouses are constructed on foundations, if your ground is unstable, you might want one. Think about how you're going to fasten it to the greenhouse frame.

The first step in creating your own greenhouse is to level the area. Gardeners in cold climates should insulate their foundations and place them below the frost line if needed.

There are two types of foundation materials, poured concrete and wood, that are preferred and widely used for ground contact.

In case you build a wooden foundation or concrete, use a tape measure and make sure to make the foundation a bit larger than the greenhouse's original finished base.

If you do not want to use a foundation at all, then you can cover the floor with bricks, gravel, landscape fabric, or any material that drains easily.

3. Building a Frame for the Structure

- Frames for greenhouses are usually made from lumber, PVC pipes, aluminum, and galvanized steel.

- Although wooden greenhouses are delightful, excessive moisture and humidity can lead to rotting. For the foundation, use ground contact-rated wood and moisture and rot-resistant wood, like cedar.

- PVC pipes are lightweight, affordable, and simple to use. These work well for modest backyard greenhouses. Polyethylene sheets are used to cover the majority of PVC greenhouses. To extend the life of the frame, select PVC pipes that are resistant to UV rays.

- Aluminum is a good material to frame glass or polycarbonate panels because it doesn't rust. Use heavy gauge aluminum for a robust, well-constructed greenhouse.

- Galvanized steel is not a popular material for backyard greenhouses because it rusts easily. Commercial growers, however, frequently cover their greenhouses with polyethylene and use galvanized frames.

4. Select the Covering Material

Pick a covering that can resist harsh weather. Various materials let in varying amounts of light. Options for greenhouse coverings include:

- Glass panels used in greenhouses: These can shatter if they are struck by hail or a stray object. Tempered glass, on the other hand, breaks into tiny, less hazardous pieces rather than sharp shards like regular glass does. In addition, greenhouse glass panels are more resilient and elegant than polycarbonate or plastic sheeting. Light pours in abundantly through the glass panels. Although they might cost more upfront, they have a long lifespan. Compared to single-pane glass, double-pane glass is a better insulator.

- UV-resistant polycarbonate sheets: Compared to glass, UV-resistant polycarbonate is less expensive and more durable because it is made of thick plastic. Although it's not as heavy as glass, it offers superior insulation and heat retention. Over time, polycarbonate plastic in a greenhouse may turn yellow or cloudy, reducing the amount of light transmitted; therefore, replacement is inevitable. The majority of UV-resistant polycarbonate has a five- to ten-year warranty. Some manufacturers offer 15 years of warranty.

- Plastic sheeting: Also known as polyethylene sheeting, this low-cost, lightweight material is used for greenhouses. But, compared to polycarbonate sheets, it is less robust. Typically, a yearly replacement of the sheets is required. For optimal insulation, use two layers of the best greenhouse plastic you can afford. Be advised that plants receive less light when there are more layers.

Chapter 4

Glazing

In general, the translucent greenhouse covering—that lets light through—is referred to as greenhouse glazing. There are numerous varieties of greenhouse glazing, and each has unique qualities. The best use and limitations of each glazing are determined by its unique characteristics.

There are a lot of things to take into account when choosing glazing material. Important considerations include the material's life, strength, weight, initial cost, light transmittance, thermal conductance, maintenance needs, and flammability.

Light transmittance

The amount of sunlight that can pass through a glazing and possibly enter the greenhouse increases with its light transmittance. Light is frequently a limiting factor for photosynthesis in colder climates and throughout the winter. Consequently, it is preferable to maximize the amount of natural sunlight that enters the greenhouse, and in most cases, it is preferable for greenhouse glazing to have the highest level of light transmittance possible. There are times when the greenhouse receives more light than is ideal, such as during the summer or in southern or equatorial regions. Under these circumstances, the amount of light entering the greenhouse may be temporarily reduced by using a shade cloth or shading compound (explained in greater detail under the "Lighting" learning unit). The shading material is eliminated when the light level falls below ideal.

A glazing's ability to transmit light is not constant. Because of aging or "yellowing" of the glazing material from exposure to ultraviolet light, as well as scratches from dust and debris, glazing typically has a decrease in light transmittance with age. In addition, the percentage of light that flows through a spotless, unobstructed glazing panel that is positioned perpendicular to the light source is known as the light transmittance value.

Strength

Stronger greenhouse glazing means that it will withstand weather-related events like hail and strong winds better than weaker glazing. As a result, the likelihood of breaking and the subsequent costs of having to replace the glazing decrease with increasing strength. Nevertheless, glazing with a high strength is frequently not very flexible.

Weight

The dead load on the structure increases with the weight of the glazing material. A more robust support structure is needed to take into account the increased dead load. This leads to higher expenses and could cause the amount of light in the greenhouse to decrease because there are more obstructions from the supporting structure (trusses blocking light, for example).

Life span

A short lifespan of the glazing requires regular replacement. Consequently, the initial cost of the glazing might be lower than that of other glazing, but after multiple replacements, it might lose its appeal in favor of a glazing that has a higher initial cost but a longer lifespan.

Scratch resistance

Debris, such as dirt, dust, and other particles, can scratch the glass. Scratching can lower the amount of light inside the greenhouse by reducing the glazing's ability to transmit light. This could, therefore, necessitate more frequent glazing replacements at a higher cost.

Cost

It's important to take into account all aspects of the glazing cost. These consist of the initial cost of the glazing material, the cost of the structural support, the glazing's lifespan, and its thermal conductivity. If a glazing material is long-lasting or has low thermal conductivity, it might be more cost-effective than other glazing materials with a higher initial cost.

Common Greenhouse Glazing Materials

Glass

There are numerous varieties of glass available, such as safety, low-iron, insulated, and floated glass. There are also variations in thickness available. The typical light transmittance of standard single-layer glass used in greenhouses is between 88% and 94% when used as a single layer and roughly 77% when used as a double layer. The light transmittance of double-strength glass is roughly 88%, while that of insulated glass is roughly 78%. The glass with the lowest iron content will transmit light the most.

Because there are gaps between the glass panels, air infiltration rates are comparatively high in glass-glazed greenhouses. As a result, compared to many other glazing materials, glass typically has a higher thermal conductance.

Glass-glazed greenhouses also generally have lower relative humidity levels than greenhouses glazed with many other types of glazing because of the higher air exchange rate (depending on how well-mounted the glazing is and the "tightness" of the greenhouse). Glass has a low impact resistance but is resistant to heat, UV light, and abrasion. In order to hold the glass panels in place and support their weight, special supports are needed, which increases the cost of installing and buying glass. Glass, on the other hand, has a long lifespan—typically longer than 25 years. Due to its high weight and expense, glass is no longer used as glazing in the majority of commercial greenhouses. Nonetheless, conservatories and botanical centers frequently use safety glass.

Polyethylene Film

Due to its flexibility, polyethylene film, a common glazing used in greenhouses, can be specially adapted to quonset structures. It is inexpensive, lightweight, and simple to assemble. A single layer of standard polyethylene film typically has a light transmittance of 85% to 87%, and a double layer typically has a light transmittance of 74% to 77%. Polyethylene film has a high thermal conductivity. However, it differs depending on the brand (e.g., Tufflite, Standard UV, Tufflite Dripless, Fog Bloc, Sun Saver, Dura-Therm) and whether it is single or double-layered.

Polyethylene glazing has a shorter lifespan than other glazing. Polyethylene needs to be replaced every two to three years if additives are not used. This is due to its high susceptibility to UV light degradation. However, polyethylene glazing may only last three to five years if additives are added during the co-extrusion process to increase the material's resistance to UV light.

Polyethylene film has a high thermal conductance, as was already mentioned. On the other hand, the internal layer of certain polyethylene film brands has an IR (infrared) inhibitor added to it, which lowers heat loss through the glazing.

The issue of condensation and dripping is another concern with polyethylene windows. On the surface of the polyethylene film inside the greenhouse, water vapor tends to condense due to the temperature differential between the interior and exterior air. Water tends to bead and gather on the surface of the film due to its high hydrophobicity until sufficiently large droplets form and fall from the glazing onto the plant materials below. The plants may become more susceptible to disease as a result of the water dripping from the windowpane. An additive that functions as a wetting agent can be mixed into the film or sprayed on it, like SunClear. As a result, water cannot bead and can instead form smaller droplets that fall through the glazing and onto the floor.

Fiberglass Reinforced Polyester

Panels made of fiberglass-reinforced polyester (FRP)—such as Excelite and Lascolite—are reasonably inexpensive, lightweight, and strong. Usually corrugated, the panels are stiff. Up to 90% of light can pass through new single panels, and 60% to 80% of light can pass through double panels. Panels can be quickly fastened with screws and rivets to wooden or metal frames. FRP is very vulnerable to UV deterioration, though. Exposure to ultraviolet light results in a decrease in light transmittance and yellowing of the panels (after just a year or two for untreated panels). To reduce yellowing and lengthen their lifespan, new varieties of FRP are given a UV inhibitor treatment.

Acrylic

There are many different kinds of acrylic panels. They could be either single Plexiglass panels or bi-wall Exolite panels. It is possible to alter the thickness of the material itself, the thickness of the panel as a whole (and hence the airspace), and the spacing between the flutes, which are the supporting cross sections inside the panels. Strength, flexibility, heat conductance, light transmittance, weight, and cost are all impacted by these panel modifications.

Polycarbonate

Similar to acrylic panels, polycarbonate panels can have different shapes and sizes. The panels can have crisscrossed supports, be single (like Dynaglass, Lexan Corrugated, and Macrolux Corrugated), bi-wall (like Macrolux, Polygal, Lexan Dripgard, and Lexan Thermoclear), or triple-wall configurations. The most widely used panel types for greenhouse glazing are single and bi-wall panels. It is possible to alter the thickness of the material itself, the thickness of the panel as a whole (and hence the airspace), and the spacing between the flutes, which are the supporting cross sections inside the panels. Strength, flexibility, heat conductance, light transmittance, weight, and cost are all impacted by these panel modifications.

Maintenance of Greenhouse Glazing

Washing greenhouse glazing on both the inside and the outside is a good idea because dirt and debris can reduce transmittance and cause scratches on the glazing in addition to blocking light. Replace any glazing that has reached the end of its useful life and has started to turn yellow. Before the heating season begins, the glazing and mounting should be checked to make sure there are no gaps between the panels, which could allow a considerable amount of heat to escape the greenhouse.

Chapter 5

Ensuring Proper Lighting

Installing grow lights is less complicated than greenhouse lighting. Your lighting options are influenced by various factors such as crop type, time of year, and sunlight availability.

For most greenhouses, six hours of full-spectrum direct light is necessary. Commercial and hobby settings usually require additional lighting. Artificial lighting can boost crop yield and encourage healthy growth. Photoperiod control lighting is also necessary for growers. The amount of light that plants receive during a 24-hour period is referred to as this type of lighting. In greenhouses, photoperiod control lighting is frequently used to accelerate or postpone flowering.

LEDs are just one of the many lighting options available to growers. Here's a quick overview of greenhouse lighting choices.

High-Pressure Sodium Lighting

Orange-red light produced by high-pressure sodium lightbulbs encouraged budding and flowering. Typically, these kinds of lighting fixtures are employed later in the growth cycle of the plant.

Energy efficiency is one benefit of high-pressure sodium lighting. Compared to incandescent lights, the bulbs have a sevenfold increase in efficiency. In order to promote healthy plant growth, the lights might also emit photosynthetic-active radiation (PAR).

High-pressure sodium lighting is inefficient in areas where lights are switched on and off frequently because it takes several minutes for it to warm up and cool down. The location of the bulbs is also very important. The distance between the lights and the plants should be between thirty and thirty-six inches.

Programmable and Fixed LED Lights

The lifespan of light-emitting diode (LED) lights is exceptionally long—roughly 50,000 hours. Since light, not heat, consumes the majority of the energy used in the bulbs, they are energy-efficient. LED lights can save up to 70% on energy when compared to high-intensity discharge (HID) lights.

Growers experience additional energy savings of up to 70% because LEDs don't require a warm-up or cool-down period. Because the bulbs don't contain mercury, recycling them is simple and doesn't require paying a hazardous material fee.

Controls can also be programmed to operate LEDs. Growers have the ability to program automatic light turn-on and off times as well as color temperature adjustments. LEDs can be used from the seedling stage to the last growth stage.

Ceramic Metal Halide Lighting

Blue light from ceramic metal halide bulbs is ideal for spaces that don't get much natural sunlight. The energy-efficient bulbs have a lifespan of approximately 8,000-15,000 hours and have up to five times the efficiency of incandescent lights.

Usually, the early stages of the plant's growth are when the lights are used. Healthy greenery is encouraged by the blue light. The drawback is that they require longer times to warm up and cool down than LEDs, which results in lower energy efficiency.

T5 Lights

T5 bulbs are the common greenhouse lighting. Compared to sodium-high-pressure bulbs and halides, fluorescent lights consume less energy. T5s are environmentally friendly and have an average lifespan of 50,000 hours. The lights can also be utilized from the seedling stage to the last growth stage.

The "T" stands for the shape of the light bulb, and the "5" for the size of the fixture. The thin tubes emit little heat, full-spectrum light, and a high lumen output. Without fear of scorching the leaves, growers can position the lights as low as six inches above the plants.

Chapter 6

Plan Proper Ventilation

An essential component of crop productivity and plant health is the greenhouse ventilation system. In addition to being essential for preserving environmental conditions, ventilation systems have a direct impact on a plant's capacity for photosynthesis, element uptake, and pollination—the process by which a plant completes its reproductive cycle. Four essential functions of greenhouse ventilation systems are air circulation, carbon dioxide/oxygen replacement, temperature control, and humidity control.

- Natural Ventilation System:

    What is a natural ventilation system? Natural ventilation is a type of ventilation system that does not use fans. Instead, the air is moved through the system by the wind and by thermal buoyancy.

    Natural greenhouse ventilation can be accomplished by:

    a. Roof Vents: These are vents at the top of the greenhouse construction that enable hot air to escape while also encouraging natural convection.

    b. Roll-Up Sidewalls: Sidewalls that roll up to allow for cross-ventilation and fresh air exchange.

    c. Windows and Doors: Opening windows and doors can also help with natural ventilation, particularly during mild weather.

- Mechanical Ventilation System:

What is a mechanical ventilation system? A mechanical ventilation system uses fans or similar mechanical devices to create sufficient airflow and air circulation. The main benefit of mechanical ventilation systems is a greater degree of control.

Mechanical ventilation options include:

a. Exhaust Fans: These strong fans remove stale air from the greenhouse, generating negative pressure that pulls in fresh air through vents and gaps.

b. Circulation Fans: These devices, also known as horizontal airflow (HAF) fans, circulate air inside the greenhouse, improving equal temperature distribution and decreasing humidity pockets.

c. Air Exchange Systems: These systems use intake and exhaust fans to keep steady airflow and a balanced growth environment.

Mechanical or Active ventilation is fantastic - especially when linked to a thermostat that turns the fans on when a certain temperature is achieved. This enables you to develop a hands-off control system, freeing up your time to focus on other farm duties.

Chapter 7

Climate Control in the Greenhouse

The greenhouse is automatically warmed to the proper temperature needed for the growth of your crops, thanks to a climate control system. The system controls and keeps an eye on fogging, humidity, shading, and a lot more.

- Using solar heating In order to catch and transform sunlight into useful heat energy, solar heating uses solar thermal collectors. Usually placed on the ground next to the greenhouse or on the rooftop.

    - Advantages of Going Green Solar Heating:

        - Reduced Carbon Emissions

        - Energy Efficient

        - Cost Saving

        - Reliable and Low-Maintenance

- Through Lights: Lighting is an essential component that keeps plant life alive. The capacity of plants to photosynthesize, develop, and grow is directly impacted by the amount of radiation. Sunlight isn't always as reliable in many climates as producers would prefer. In these circumstances, greenhouses frequently use additional grow lights to extend the day or give light on gloomy or foggy days.

Chapter 8

Growing plants in the Greenhouse

a) Seed Starting and Propagation

With greenhouses, you can regulate the humidity and temperature needed for seed propagation and the growth of young seedlings. In fact, you can start seeds in greenhouses at any time because of this controlled environment. However, if you are starting plants that you intend to transplant outdoors into gardens in the spring, then you should be starting the seeds in greenhouses six to eight weeks prior to the last anticipated frost date for your area. Most seeds should germinate best in temperatures between 70 and 80 degrees F (21-27 C), with nighttime temperatures that do not go less than 50 to 55 degrees F (10-13 degrees C). Maintaining a close eye on the temperature in your greenhouse is necessary. During the day, when the sun is shining, greenhouses are usually warm, but at night, they can get much colder. Heat mats for seedlings can be used to help maintain a constant warm soil temperature for seeds. Overheated greenhouses can be vented using fans or opening windows in the greenhouse.

Seed Starting:

Typically, individual plug trays or open flat seed trays are used in greenhouses to start seeds. Seeds are prepared based on their individual requirements; they could be stratified, scarified, or soaked for an entire night before being planted in trays inside the greenhouse. In order to facilitate thinning, watering, fertilizing, and treating diseases that affect seedlings, like damping off, the seeds are typically planted in neatly spaced rows in open flat trays.

These seedlings are then moved into individual pots or cells after they grow their first set of true leaves. There are only one or two seeds planted per cell in single-cell trays. Since plug cells hold and retain more moisture and warmth for the developing seed, many experts believe planting in plug trays is preferable to planting in open trays. Also, seedlings can be kept in plug trays for an extended period of time without their roots joining with those of their neighbors.

Plug-grown seedlings are easily removed and placed directly into the garden or container plantings. You don't have to spend a lot of cash for specialized seed-starting mixes when starting seeds in a greenhouse. Make your own all-purpose potting mix by combining perlite, peat moss, and organic material (like compost) equally. Nonetheless, it is crucial to sterilize any potting medium you use in between applications in order to eradicate pathogens that may cause damping, a disease that affects seedlings. In addition, weak, lanky stems may develop in seedlings if the greenhouse's temperature is too low, the light is not bright enough, or if they receive excessive watering.

b) Growing Plants During Cold and Warm Weather

Greenhouse Seed Starting for plants in summers:

Growers in greenhouses should be getting ready for their second growing season, which takes place in the greenhouse. In order to grow year-round, plants often need to be started early, in the late summer or early fall, when the days are still sufficiently long. This enables seedlings to reach maturity before light levels diminish and to receive adequate light. A year-round harvest is produced by the robust, healthy plants that grow in the greenhouse throughout the winter months. Seedlings that are started too late will have lower light levels and grow twisted.

In the summer, the greenhouse is a place of fresh starts. Depending on their planting strategy, farmers can begin their crops in trays or straight in the ground. The local climate and greenhouse design will play a major role in determining the seeds to be planted for the second growing season. They should be well-planned way earlier than the time of planting.

In the summer greenhouse, you can grow almost anything.

It is preferable to convert to plants in your summer greenhouse that can withstand the heat and yield a large number of crops. For your summer greenhouse garden, tomatoes, peppers, eggplant, beans, and other heat-loving plants raised in the soil from early spring may be your best bet. Grown in a warm greenhouse, they will keep producing all summer long as long as the soil is well-hydrated. Other plants that would look great in your summer greenhouse include the following:

- Greens are also always necessary. Yes, they can grow well in the open air during nice weather, but they could grow considerably more quickly if they are planted in a greenhouse.

- A for salad, every two weeks, you can plant them, and for that, you do not need the salad seedlings. It grows quickly as well, given that the greenhouse maintains the proper temperature. A word of caution: water directly beneath the roots, avoiding the leaves; loosen the soil and weed it. Avoid planting salad in a mound. Types include watercress, cabbage, asparagus salad, romaine lettuce, and cabbage.

- Strawberries enjoy warmth, light, and loose soil. For this reason, adding peat to the soil is a good idea. It grows in a single tier on the open ground. The "block method" can be applied in the greenhouse. After being planted in garden boxes, strawberries are stacked in multiple layers on the rack. Artificial LED growth light is what we recommend in case you are not getting enough light during the day. Install an irrigation system to make it easy for you to take care of them. You can harvest more than 100 pounds in the greenhouse if the right conditions are met.

- Because of its highly erratic culture, dill is ideal for summer greenhouse cultivation. It requires well-moisturized soil, good lighting, and a minimum temperature of 60 degrees Fahrenheit (15 degrees Celsius). Dill takes around two months to grow. However, if you grow 10 square feet of this green seasoning and periodically trim off the green tops, you can harvest up to 5 pounds of it, so it's worth it.

- Parsley. It can be grown by planting roots down to a depth of 6 inches, with a 2-inch space between each root, or by scattering seeds and waiting for them to germinate before covering them with a damp cloth and planting them in the ground. With the second method, you can get more than 3 pounds of harvest per 10 square feet of growing space during the summer. Although the first approach is a little more challenging, it would be more resilient. Simple maintenance for parsley involves weeding, watering, and adequate ventilation.

- While planting, we have to make sure that the soil is leveled, fertilized, and loosened. Place the seedlings one inch apart from one another. As for spouts, to encourage their rapid growth, we can cover them with thatch straws that are mixed with manure.

c) Growing Vegetables, Herbs, and Fruits

Vegetables:

A greenhouse is suitable for growing vegetables in a number of ways:

1. Planning

- Calculate the floor space and greenhouse beds needed for grow bags. Verify that you have enough space to grow every summer greenhouse crop you intend to grow.

- Plenty of space should be provided by benches for seedlings, a lot of which will be moved outdoors when summer greenhouse crops require that area.

- A quickly maturing crop is planted in space that is freed up after one main crop is harvested and before the next is sown or planted.

- Salad leaves can be sown as a catch crop in the greenhouse borders prior to the need for greenhouse space for the summer crops.

2. Sowing Seed Indoors

- Make use of spotless pots and trays, fresh seed free of peat, or multipurpose compost.

- Stick to the directions on the seed packet.

- Seeds will develop on a sunny windowsill indoors.

A portable heated propagator is a structure with a vented lid and a thermostat that is responsible for controlling the temperature continuously and adjustably. It creates a warm, humid environment that promotes rapid rooting in cuttings and the germination of seeds.

A propagator is a lightweight, transportable structure made of plastic that can have a vented or unvented lid to create a humid, a little warmer environment. It is helpful for root cuttings and seeds to germinate. It might have a thermostatically controlled temperature system.

3. Growing on

- After germinating, seedlings require a bright, frost-free surface to grow on; an unheated greenhouse might not be warm enough until April.

- If you would like to create a good growing environment for tender plants, think about fleecing and heating a section of your greenhouse that is divided.

- Keep an eye on the weather forecast and be ready to cover young plants with fleece on chilly nights or give them extra heat as needed.

4. Planting

- Once the protected crops are strong and well-rooted, plant them in their final positions.

- Plant in growing bags, containers, or greenhouse borders.

- To make a shallow bed, cut off a long panel from the top of the growing bags before using them for autumn salads.

- Make sure melons and cucumber, which are climbing plants, have enough support and tying.

Cordon planting is a common practice for growing apples, pears, gooseberries, tomatoes, red and white currants, and especially in areas with limited space. You can grow sweet peas as cordons to get large flowers for display.

5. Summer Maintenance

- Install irrigation or check watering every day; improper watering can lead to issues like tomato blossom end rot.

- Open doors and vents to let fresh air into greenhouses on warm days; automated ventilation is best.

- For heat-loving vegetables like cucumbers and okra, vents can be left closed, but damping down is necessary to increase humidity. As an alternative, use fleece or transparent plastic to divide off a portion of the greenhouse.

- There will need to be some shading; this should be added gradually because it will initially slow down growth.

- Use sticky yellow traps to identify pests early on. Then, biological controls can be ordered right away.

- Regularly tie new growth into supports and pinch out cordon tomato side shoots.

6. Winter Maintenance

- Keep structures clean, especially the glazing.

- If it is a heated greenhouse, make sure that a thermostat is there to maintain the night-time temperature.

- Have a thermometer to monitor the conditions

Problems

When cultivating greenhouse borders, remove the soil every three to five years and replace it with high-quality garden loam or purchased topsoil to prevent the accumulation of soil-borne diseases. Where soil problems are suspected, grafted plants (available now include tomatoes and aborigines) may be helpful.

If the soil appears questionable, another option is large pots and grow bags. Cover the soil with a white plastic sheet and set pots or bags on top of it. If there are no diseases present, potting mediums, including grow bags, can be reused at least once for different crops.

A few diseases to be aware of are powdery mildew, grey mold, and damping off. Glasshouse whiteflies, glasshouse leafhoppers, and glasshouse red spider mites are common pests under glass.

Herbs

Any dish can benefit from the exciting depth of flavor that fresh herbs can provide, as well as the ability to evoke the cuisine of a specific country. Should you frequently use herbs in your cooking, you may want to consider cultivating your own to guarantee a steady supply.

Compared to herbs you would pick yourself from your garden, herbs bought from a store are far less fresh and frequently come in unnecessary packaging that harms the environment. Purchasing fresh herbs from a store or greengrocer on a regular basis can be costly as well. Since a single bunch of herbs is about the same price as a packet of seeds, it is far more cost-effective to just grow your own.

d) Growing Herbs in a Greenhouse

While most herbs grow happily planted directly in your garden's soil, they can also be grown and thrive in a greenhouse.

You can reliably and with additional protection grow almost any herb in your own garden, whether you have a full-size glasshouse or a small greenhouse.

Grow Herbs Throughout the Year

You may increase the length of your herbs' natural growing season by using a greenhouse to help control temperature and hold on to soil moisture. Thus, you can still enjoy fresh herbs when they're not in season.

Protection From Uncertain Weather

Delicate herbs are well-protected in the safe environment of a greenhouse, which promotes healthy growth and reduces plant loss. It protects them from harsh weather that could cause your herbs to struggle or perish, such as severe rain, hail, freezing temperatures, and excessive sunshine.

Protection From Pests:

Many animals in your garden could be interested in enjoying your recently harvested herbs. In addition to keeping your herbs pesticide-free, the physical barrier provided by a greenhouse makes it much easier to ward off rodents, caterpillars, and snails.

What Herbs Grow Well in a Greenhouse?

Any herb can be grown in a greenhouse. Some of the common ones include:

- Basil

- Chives

- Coriander

- Dill

- Parsley

Hardier herbs can also be grown for a longer period of time by being kept in a greenhouse. They include:

- Oregano

- Rosemary

- Sage

- Thyme

Tips For Growing Herbs in a Greenhouse

As with any plant, you should keep a close eye on your herbs to make sure they thrive all year round.

Don't Over-Water

While it's crucial to water your herbs until the soil is moist, letting them dry out a little bit between watering promotes the growth of robust root systems.

Pinch and Prune

To keep your herbs from going to seed and to extend their growing season, pinch off any flowering stems. Once the plant is a few inches tall, you can also trim the small leaves at the top so that the larger leaves at the bottom can form a strong base.

Let Them Breathe

Plant growth requires a lot of fresh air to be healthy. In order to avoid stale air, which can encourage the growth of fungi in the warm, humid atmosphere of your greenhouse, make sure to regularly ventilate it.

Fruits

To begin with, not every species can be grown in a greenhouse. Certain fruit trees are better suited for greenhouse cultivation than others, whether it's because of their size or other requirements.

Grapes

A common misconception is that growing grapevines require constant heat, but you can grow your own grapes if you choose the right variety with care. Of course, the owner of a heated greenhouse has plenty of options, and varieties like "Black Hamburger" or "Buckland Sweetwater" are perfect for cooler growing conditions. For vines to thrive, the soil must be open and free-draining; this is a crucial part of their cultivation, as soggy soils can have detrimental effects. Sturdy canes are usually the best option for supporting young plants until they become established, and as they grow, new spring growths can be trained to support them. The side shoots that are produced in the summer should be pruned in the following winter, cutting them back to their last bud, a technique known as "spur" pruning, in order to encourage a good crop. Vines bear fruit on the new growth that follows each year.

To keep the plants at their best, they will require some care throughout the year. They will need to be fed with high-quality fertilizer and well-watered to guarantee that enough water reaches all of the roots. To ensure that every grape has enough room to grow to a good size, the fruits must be thinned as they develop. This process is typically done with long, pointed-end scissors, working upward from the bottom of the bunch to remove the smaller grapes.

Unfortunately, aphids, red spider mites, and scale insects are just a few of the pests that affect vines. As such, it's important to keep a close eye out for any problems and act quickly to address them when they arise. Mildew can also affect the fruits. However, even the smallest greenhouse can produce a few bunches of grapes grown in-house with a little attention and work.

Peaches and Nectarines:

You can grow nectarines and peaches in cool greenhouses or in unheated ones. Just like with vines, choosing the appropriate variety is essential to good fruit production. "Peregrine" peaches are a great option since they ripen in August and are self-fertile. "Hale's Early" peaches, while still a great choice, require close proximity to another variety for successful pollination. Nectarines such as "Humboldt" and "Pine Apple" are excellent choices for a cool or unheated greenhouse because they ripen later in the season and yield fruit in September and frequently even into October.

It takes some time and work to keep these trees healthy, but the benefits are well worth the investment. In contrast to vines, fruit trees typically bear fruit on the wood from the previous year. This allows the fruits to grow and makes picking them easier. Pruning and training the tree to keep a fan-shaped canopy is common.

Because the roots of peaches and nectarines cultivated in greenhouses are often rather compact, they require regular watering during the growing season and fertilization after the tree has established itself. The plants in the greenhouse require assistance from pollinating insects, which can only be provided by using a small, fine brush to move pollen from one flower to another.

Since not all pollen ripens at the same time, this must be done daily throughout their flowering period. To increase the likelihood of success, the greenhouse's humidity should also be increased. There may occasionally be a need to thin the fruits as they grow. When the fruits are roughly the size of walnuts, this task should be completed, thinning out so that about two fruits can develop per foot of branch.

Citrus Fruits

It may come as a surprise to learn that you can grow tangerines, oranges, and lemons in a greenhouse. As long as their surroundings remain above 4 degrees C (40 degrees F), they will happily survive the winter. The pips only need a temperature of about 13 degrees C (55 degrees F) to germinate. Even though starting them from seed is the simplest method to create your own tree, it can take up to ten years for them to start bearing fruit on their own. Taking appropriate cuttings should enable fruiting plants to be produced significantly faster if faster propagation is desired.

Fruit cultivation in a greenhouse is not without its challenges. In the end, nothing compares to being able to stroll out to the garden and pick your own produce. It may take some work to get the conditions just right, protect against pests and disease, and even raise the plants correctly.

Chapter 9

Dealing with Pests and Plant Diseases

Plant pests have a direct and significant impact on the potential greenhouse vegetable producer. Even as plants are being ready for shipping, pest and disease problems can have a disastrous effect on greenhouse productivity. When it comes to pests and diseases, prevention is cheaper than treatment.

Necessary procedures for any farm having a greenhouse:

- To guarantee early detection and accurate problem identification, do checks twice a week during the summer and once a week during the winter. You should also be able to accurately identify diseases and pests or have them diagnosed for you.

- Create action points that indicate the appropriate times to apply chemical, biological, whole-crop, and hot-spot treatments.

- Create a "clean" zone surrounding the greenhouse, keeping it isolated from the farm's "outside" area, which includes roadways and residences. Prior to entering the clean zone, everything needs to be cleaned.

- To monitor and control the flow of people, cars, plants, and materials into the "clean" zone, use check and control points such as gates, signs, and wash bays.

- People moving around the farm and into other greenhouses are popular methods to introduce pests and diseases, so be sure that staff members and guests who have visited other greenhouses change their clothes and shoes before entering your greenhouse.

- Before putting any seedlings into a clean greenhouse, make sure they are free of pests and diseases.

- Keep a clean, clear buffer zone around each greenhouse that is 5 to 10 meters wide. This will cut down on the amount of pests that enter from nearby paddocks, minimize the need for spraying, and keep giving farms access in all weather conditions.

- To lessen the chance of pests and diseases lingering from the previous harvest, make sure the greenhouse is thoroughly cleaned and sanitized before starting a new crop. When cleaning, make sure you have a detailed, step-by-step work method.

- One of the main sources of pests and diseases is weeds; therefore, keep the greenhouse and the surrounding agricultural area weed-free.

- As soon as a crop is done, remove and dispose of crop trash into a cart or bin outside the "clean" zone and away from the greenhouse.

Chapter 10

Fertilizers to be used in the Greenhouse

Most greenhouse plants require fertilizing since containers rarely supply enough nutrients for long-term development.

Carbon (C), hydrogen (H), oxygen (O), phosphorus (P), potassium (K), nitrogen (N), sulfur (S), calcium (Ca), magnesium (Mg), iron (Fe), boron (B), manganese (Mn), copper (Cu), zinc (Zn), molybdenum (Mo), and chlorine (Cl) are the 17 elements necessary by all plants. Nickel (Ni) was recently shown to be a vital nutrient for plants (2004). The elements C, H, and O are mostly provided by air (carbon dioxide (CO_2) and oxygen), as well as water (H_2O). The remaining 14 elements, known as mineral nutrients, are obtained from a variety of sources.

Nitrogen, phosphorus, and potassium are the three basic plant nutrients included in standard fertilizer. Secondary plant nutrients such as calcium, sulfur, and magnesium are also found in fertilizers. Boron, manganese, iron, zinc, copper, and molybdenum are some of the micronutrients contained in fertilizers. Fertilizers are classified into two types: inorganic fertilizers and organic fertilizers. Organic fertilizers are obtained from plants and animals, whereas inorganic fertilizers are extracted or created from non-living sources. While precise comparisons between these two sources are impossible, inorganic fertilizers are the favored choice in most greenhouse operations.

The percentage of each is shown on the fertilizer package by a numerical ratio, with the order always being N to P to K. A ratio of 10-5-5, for example, indicates that the fertilizer has twice as much nitrogen (10%) as phosphorus (5%) and potassium (5%). Fertilize plants only during their active growth season, and always follow the package directions to ensure you apply the correct quantity. Many people believe that if a little fertilizer is beneficial, a lot must be much better, but this is not the case. In reality, too much fertilizer, especially synthetic fertilizer, can be harmful to plants.

Chapter 11

Hydroponic and Aquaponic Cultivation

Hydroponic Cultivation Techniques

Hydroponics is a production method where plants are raised in various growing mediums and hung in nutrient-rich water instead of soil.

In soil production, plants only have a limited time to absorb nutrients before they sink through the substrate, which wastes valuable food and diminishes growth. Inside a hydroponic greenhouse, improved nutrient delivery and superior growing conditions produce a cohesive environment for plants.

Benefits Of Hydroponics

- No weeding and minimal or no pesticide use

- Easier pest control

- Quicker crop turns

- Higher density crop

Types of Hydroponic Systems

There are many different types of hydroponic growing systems. While these utilize the same principles, they use different techniques, each with its own pros and cons.

a) Deep Water Culture (DWC):

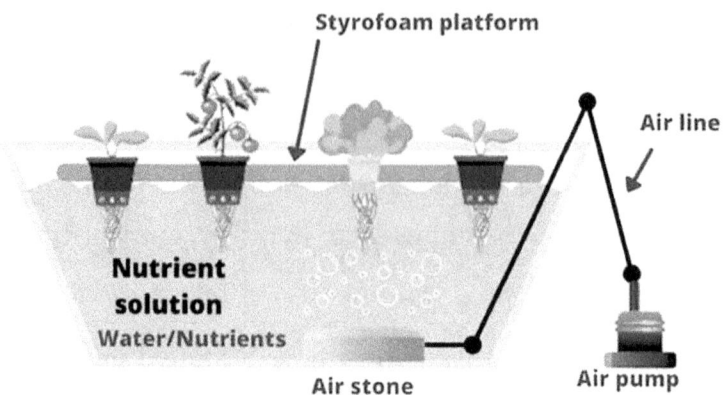

In a DWC system, the plant's roots are suspended in a solution of water and nutrients that is well-oxygenated. DWC is ideal for small plants such as lettuce and some aromatic plants.

b) Wick systems

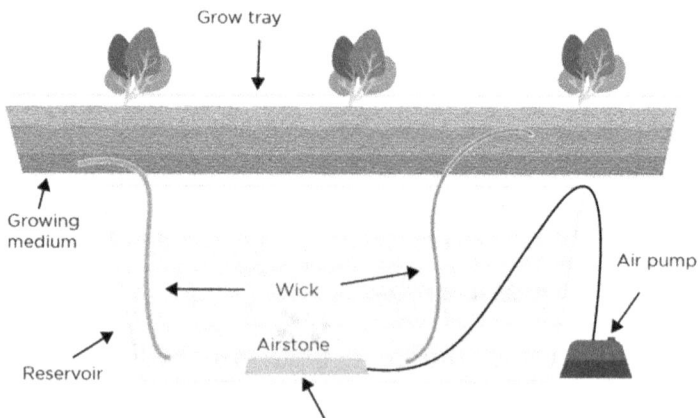

The Wick system is perhaps the most straightforward in terms of both design and function. In contrast to other hydroponic systems, this one doesn't need an electric pump to move nutrients and water to the plant roots. A string would be enough as the wick! From the roots to a source of nutrients and water, a wick stretches downward. Normally, the plants are planted on a substrate made of wool, sand, or pebbles. This technique is less effective when used on a broad scale and functions best with individual plants.

c) Ebb and Flow

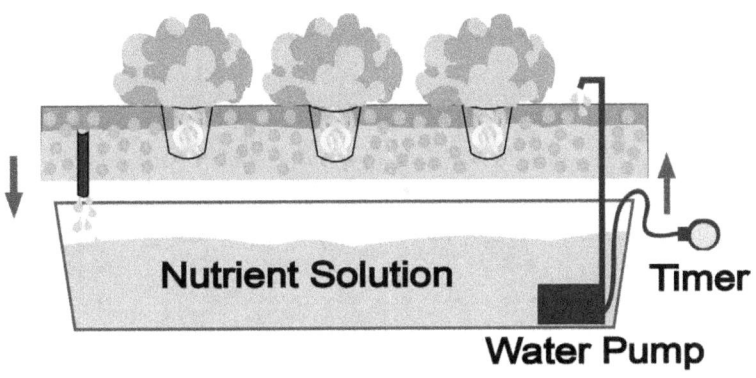

Ebb and flow, also known as flood and drain, is one of the most widely known hydroponic systems. The system is comparatively easy to set - up, too simple to operate, and adaptable enough to successfully grow almost any kind of plant. The idea is extremely straightforward: after placing plants in a tray, nutrient-rich water is pumped into the tray from a reservoir that is positioned underneath the grow tray. Water runs through the roots and the base of the pots before returning to the reservoir via the drain tube. Before flooding the grow tray once more, the plant roots are given some time to dry out and get oxygen. Usually, a timer-connected submersible pump is used for this operation. Depending on the size and variety of plants, the timer is configured to turn on many times a day.

d) Drip systems:

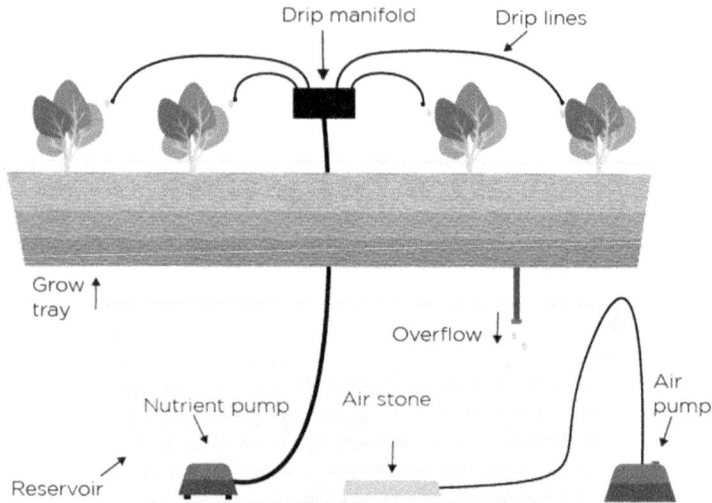

The most widely used hydroponic system. It is one of the most frequently applied types of hydroponic systems across the world, both for home gardeners and commercial producers alike. Hydroponic drip systems could be built in a variety of ways, ranging from simplistic to large networks. However, they are extremely helpful for bigger plants that require a lot of root area. This is because flooding the system does not require enormous amounts of water. In a drip hydroponic system, water and nutrients are gradually delivered to the plant roots by drip irrigation. Water is delivered in gentle drips to keep the plant roots wet but not overwatered. This approach lowers water evaporation since the water drips slowly, as well as water waste due to contamination or runoff. There is control over how much water is provided to the plant with a drip hydroponic system.

e) Nutrient Film Technique (NFT):

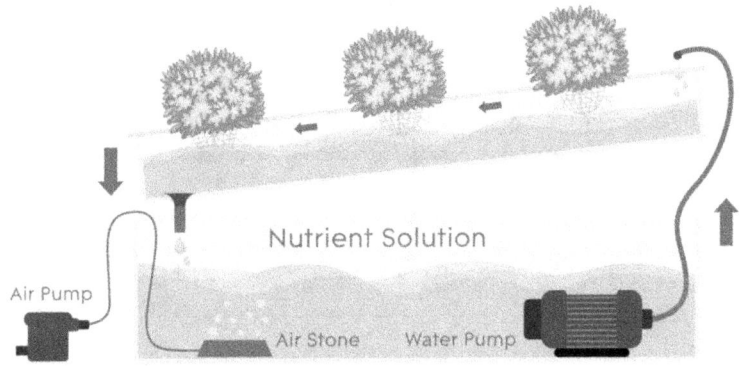

This technique is similar to the flood and drain system, except the water-nutrient solution circulates in a continuous loop around the reservoir and growth tray rather than with periodic pauses. The grow tray is angled to enable water to flow down toward the drainpipe, and a new solution is continually fed into the tube's upper end. While this approach does not require a timer, it does require two pumps and an air stone to oxygenate the water, as well as tubes or channels to set your plants in.

f) Aeroponic Cultivation Techniques:

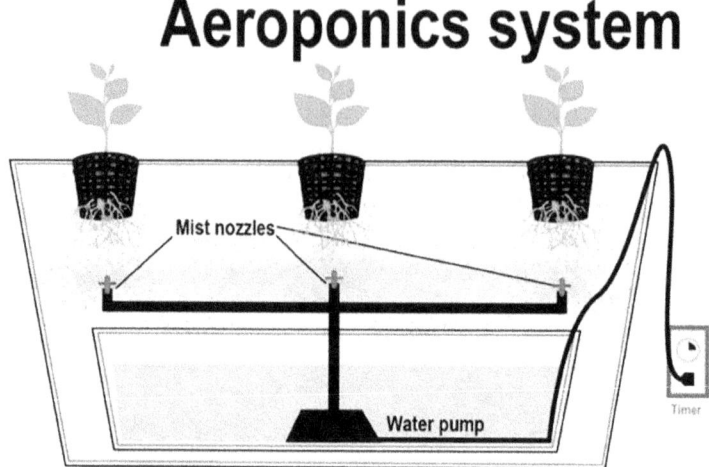

Aeroponics is a method of growing plants without soil. A nutrient-rich mist is used to irrigate roots that are suspended in the air. Large vertical grow racks are used to hold plants. Nitrogen, phosphorus, and potassium are among the essential organic liquid nutrients that are added to a large reservoir of water. Using a sprayer to create tiny water droplets, fogging or misting technology delivers water to the exposed roots. An extended duration of water droplet suspension in the air causes a higher rate of evaporation and a drier canopy of plants.

Chapter 12

Maintaining your greenhouse

Clean Your Greenhouse Effectively

It is crucial that you clean your greenhouse thoroughly because a dirty greenhouse makes it easier for diseases and pests to spread and makes it more difficult for the sun's valuable light to enter.

Stick to these tasks when you come to perform a deep clean on your greenhouse:

Interior:

1. Take everything out of the greenhouse, being especially careful about what is beneath your pots and plants. It's possible to discover insects, eggs, or slugs that need to be removed.

2. To save time when sweeping up all of the debris from your greenhouse at the end, thoroughly clean and wipe down all sides of all removed pots. Work from the top down.

3. Fill a bucket with warm, soapy water and start scrubbing the greenhouse from top to bottom. Scrubbing areas that are heavily covered in moss or anything similar should receive a lot of attention.

4. After giving the entire greenhouse a thorough cleaning, grab an old toothbrush and start cleaning the harder-to-reach places, making sure to give them a thorough cleaning.

5. After cleaning the greenhouse completely, make sure to give the floor a thorough scrub and a last rinse with clean water.

6. Give the greenhouse time to dry completely; don't rush it; this could take a day or two!

You might want to use a reliable product to disinfect the greenhouse after it has dried. This will guarantee that no illnesses or pests return for a considerable amount of time. When choosing a disinfectant, be cautious to stay away from any potentially dangerous chemicals, especially if you grow organically or have animals that roam freely.

Exterior:

1. Clear the area around the furniture, flowers, decorations, and other items.

2. Get rid of any weeds that have established a home near the building's foundation and base. The sides of the greenhouse that are less accessible, like those that are next to a wall or fence, should receive special attention. Cleaning will be simpler as a result, and it will also help locate any trouble spots, like any cracks or openings where mice or other pests could get in.

3. Turn off the auto vent openers. This will prevent errant vents from opening in the sun while you are cleaning the roof, which will make the task much easier.

4. Use a soft-bristled brush to remove any leaves from the greenhouse's roof and clean the gutters. If you want to store your Rhino blinds indoors for the winter, now might be a good time to take them down.

5. After removing any large debris, start using soapy water to scrub the structure. To reach all the way up the ridges, use a long mop with an extendable handle.

6. To ensure that the glass is as clean as possible, we advise running a fresh bucket of soapy water over the structure once more.

7. To prevent smears, you might want to buff the glass after cleaning it with a gentle cloth or cotton t-shirt.

NB: Avoid using acidic cleaning products, as this will corrode the aluminum frame.

You'll be happy to learn that cleaning Rhino Tuff safety glass is significantly simpler than cleaning polycarbonate and horticultural glass. This is primarily because of its stronger makeup, which means that when cleaning, you don't have to worry about applying too much pressure to the panes. It's also important to note that there are smooth, easy-to-use flat surfaces instead of awkward grooves, as you might find with polycarbonate!

How often should you clean your greenhouse?

Cleaning your greenhouse's entire structure should ideally be done at least once a year. Since cleaning is much simpler and takes less time during the winter months, when your greenhouse is probably empty, we would advise cleaning your greenhouse during those times. Before sowing begins, early spring is a good time for a clean. If you've cleaned well during the winter, you'll have less work to do in the spring.

Eliminate powdery mildew in greenhouses

Powdery mildew is one of the most prevalent types of fungal plant disease. The disease's obvious symptoms include the appearance of powdery white or grey patches on the leaf's surface. Thankfully, powdery mildew rarely causes plants to die; instead, it usually appears much worse than it is. A crop that is particularly vulnerable to this problem is cauliflower.

That being said, if powdery mildew does manage to get hold of your crops, your plants will look much worse, and their growth will slow down as they battle the illness. For organic gardeners, in particular, the simplest solutions are frequently the most effective ones, so try easing the symptoms by mixing baking soda and liquid soap. Because baking soda can corrode aluminum, be careful to wipe it off as soon as you find any on your staging or greenhouse.

You can minimize the likelihood of powdery mildew harming your greenhouse plants by taking the following preventive measures:

- Prevent powdery mildew by keeping plants well apart from one another. Proper air circulation is essential for this.

- Steer clear of over-fertilizing your plants; slow-release fertilizers are better because they allow for more controlled growth.

- Reduce the amount of shade your plants receive. In a greenhouse, this shouldn't be an issue, but make sure your plants get at least six hours of sunlight every day.

How To Get Rid of Greenfly in Greenhouses

You're not alone if greenflies have taken over your greenhouse. The greenfly has gained a bad reputation for being one of the most bothersome garden pests because of its abundance; once one enters your greenhouse, many more will follow quickly.

Fortunately, you have lots of options for getting rid of these annoying bugs in your greenhouse:

- Smash them off: All you need to do is take a hosepipe and blast the offending plants. If you're working with sensitive plants and crops, this method can be difficult, but it shouldn't be written off. Since this is where greenflies are most likely to hide, make sure to look beneath the leaves!

- Sticky traps: Using a few homemade or store-bought sticky traps is an easy and affordable method of eliminating greenflies from your greenhouse. Just take some yellow cards, smear them with petroleum jelly or washing liquid to make them sticky, and hang the traps close to the plants that are most impacted. Greenflies will be drawn to these traps, and you won't have to worry about them bothering your greenhouse anymore!

- Insect repellent: Use caution when applying this technique since some crops may suffer more damage if they come into contact with it. In a greenhouse, always research the insect repellent you use because you might discover that it negatively impacts your plants.

How to Keep Mice Out of Your Greenhouse

It goes without saying that rodents may do a great deal of damage to your greenhouse. They cause mischief by biting through plastic pots, wood, and anything else they can get their hands on, not to mention devouring your seedlings!

You can take a few easy precautions to keep mice out of your greenhouse, but you really need to be on the lookout for rodent activity in your garden and greenhouse at all times. First of all, make sure to block off any open spaces in your greenhouse, such as cracked glass or malfunctioning louvers. It's that easy: if you prevent mice from entering your property, you shouldn't have any problems with them!

We advise cleaning up your greenhouse so that mice have nowhere to hide if there are no easy access points for them to enter. Rodents can easily hide among the weeds and overgrown grass in your greenhouse, so make sure to keep everything tidy. It should go without saying that you should not store pet food near your greenhouse as this will draw mice and other rodents.

How to Get Rid of Fungus Gnats in Greenhouses

Fungus gnats are a general nuisance in your greenhouse and are known to spread fungal pathogens like Botrytis, even though they won't actively eat your crops. You're justified in wanting to get rid of the fungus gnats in your greenhouse.

Worrying about keeping your greenhouse safe from these pests is a lot like being concerned about your greenhouse being safe from mice because you'll need to keep your greenhouse clean and weed-free. Since fungus gnats are drawn to dead and decaying plant matter, it's critical to regularly clean your greenhouse thoroughly and keep an eye out for overgrown and decaying vegetation. In addition, keep in mind that fungus gnats are drawn to an abundance of moisture, so avoid overwatering plants.

If you have a persistent issue with fungus gnats, you might want to think about moving your compost bin if you keep one in the greenhouse.

Use biological control techniques, such as nematodes, if these preventative tips aren't working to get rid of the pests. Since nematodes are safe for your plants and very effective against fungus gnats, we prefer to use them.

Preserving Your Greenhouse

High winds & storms:

It's crucial to get your greenhouse ready for strong winds and storms as soon as you can, preferably a few days ahead of the predicted weather. Making your greenhouse as airtight as possible and securing the structure are the most crucial things to do. Make sure to secure everything in your greenhouse because it can cause a lot of problems if the wind gets inside.

Keeping Your Greenhouse Insulated in the Winter

Maintaining the highest possible temperature in your greenhouse throughout the year is crucial, particularly if you intend to keep plants inside during the winter. Many greenhouse owners find that bubble wrapping their greenhouse is an effective way to retain a significant amount of heat inside the structure.

The layer of insulation is what the bubble wrap is there for. But it's best to use horticultural bubble wrap rather than the standard packaging option.

It is ideal to complete this after the yearly cleaning. Ensure that every window has been cleaned, all crevices and corners cleared, and general cleaning of debris, bugs, and illness has been completed. After cleaning, let the greenhouse air out and dry. It's critical to lower the interior moisture content because, once the greenhouse is wrapped, you won't have easy access to the vents and louvers.

It's also crucial to keep the doors open frequently to avoid the accumulation of everything we've just been teaching you to keep out because bubble wrapping the greenhouse hinders air circulation!

What to do if the glass breaks?

Nothing is more annoying than having a greenhouse glass break; it poses a serious risk to everyone using your garden area and is a major inconvenience. Depending on the kind of greenhouse glazing you have installed, you may need to take a different approach to dealing with the breakage.

Because horticultural glass has sharp edges, it must be disposed of carefully, so handle it with caution. In order to replace the pane in your greenhouse, you will need to take out all of the clips that hold the glass in place and carefully remove the pane, covering the hole left by its removal.

Chapter 13

Managing Cost

A successful greenhouse has to decide on and set rates that will allow it to turn a profit while also covering all of its expenses. For a greenhouse business, choosing a price is one of the most challenging marketing-related decisions. Prices need to be defined at a level that clearly reflects customers' willingness to pay.

Growers are primarily concerned with three issues: labor availability, profitability, and competition. Labor is the largest cost, followed by energy.

In order to combat the shortage of laborers, greenhouse production facilities are increasingly utilizing technology, machinery, and automation.

To make sure they have the labor force to plant, grow, and transport their goods within the restricted shipment windows. Many greenhouse businesses are also taking H2A labor programs into account.

Strategies growers use to save labor costs:

- Product selection and size are being streamlined.

- Sticking machine for robotic vegetative cutting

- Robotic transplantation

- Grading and sorting system that is automated

- Increasing the usage of conveyors

- To reduce by hand pinching, using a trimming machine or chemical pinches.

- Irrigation booms reduce the need for manually watering

- LEAN-flow techniques (progressive sticking, supermarket-style shipping)

Energy is the highest overhead cost after labor. In an industry in which profit margins are shrinking and there is a greater need for sustainable production, more focus is being placed on producing greenhouse crops in an energy-efficient and environmentally friendly beneficial manner.

Growers can enhance their use of energy in a more efficient manner by:

- Installing infrared reflective (IR) poly.

- Installing high-performance heaters

- Installing shade and/or energy curtains to reduce heat

- Replacing lighting with more energy-efficient options like LEDs.

- Adding solar photovoltaic systems

- Insulating side and end walls for efficiency

- Using in-floor heating to decrease air temperature and heat loss

- Utilizing horizontal air flow (HAF) fans to circulate air

- Using environmental control systems for automatic ventilation, heating, and lighting

Chapter 14

Making the Most of the Greenhouse (Profit)

Not only are greenhouses a common sight at botanical gardens, but They are also a year-round method of growing vegetables. You and your family could earn money from those veggies. You can earn money by growing vegetables in your greenhouse that aren't normally available in the winter or that aren't available in a certain area and selling them directly to consumers at farmer's markets.

1. Think about your business plan. Think about the cost and distance involved when you need to deliver your veggies from your greenhouse to farmers' markets and other locations. These expenses reduce your total income. Recognize that location is crucial if you choose to have people come to you. Make sure that a large market or populated area is not far from your greenhouse.

2. Selecting the crops you will grow in your greenhouse requires careful thought and research. Grow vegetables that aren't readily available in your area during specific seasons of the year to optimize your profits. When you can provide fresh vegetables even when they're out of season in your area, greenhouse growers have a significant advantage. They can grow at any time of the year.

3. Make sure your product is of high quality. Producing a high-quality crop on a regular basis is crucial to turning a profit

from your greenhouse. If the vegetables are of the same quality as what they can get at a grocery store, people won't make the extra effort to visit your greenhouse or purchase your product. Utilize the best seeds and technology available, even at a higher cost, to guarantee repeat business from your clients.

4. Keep an eye on your greenhouse's costs. You might be thrilled with all the extra money you have after your first crop. However, take into account how much of that money was spent on crop cultivation, crop selling costs, and greenhouse upkeep. Pay close attention to how much money you earn and how much you take in. Success lies in understanding the differences.

Chapter 15

Avoid Common Mistakes

The uncontrollable nature of the elements can be a cause of frustration for gardeners engaged in outdoor gardening.

Although there isn't much we can do to alter the natural conditions of our gardens, we can try to work with them.

We can adjust the light, temperature, and humidity to the point where you feel like an expert gardener.

However, as numerous mythological tales have demonstrated, gods are also capable of making mistakes.

In order to help you make the most of your greenhouse's capacity to grow plants, we'll go over some typical mistakes that gardeners may make.

1. Choosing a bad location:

   - When searching your property for possible greenhouse sites, there are a few key things to think about.

   - Try to locate the structure south of your home to optimize the amount of year-round sunlight it receives in the Northern Hemisphere.

   - The next best regions are in the east or southeast, then the southwest and west. If at all feasible, try to locate it south of your house.

- It makes sense to locate it somewhat close to your house so you won't have to walk or run your utilities very far. If you need a larger structure for gardening, a detached greenhouse in the yard works well, but an attached greenhouse is as close as it gets.

- Furthermore, it might be a good idea to keep your human and plant children apart. One could even serve as a calming haven in case the other is making you feel anxious.

Regarding trees in the vicinity, you have two choices. Make sure there are no nearby trees that would prevent your greenhouse from receiving as much sunlight as possible.

However, a deciduous tree placed properly can provide afternoon shade during the sweltering summer months while still letting light pass through its leafless canopy in the winter.

Make sure the structure rests on level ground wherever you decide to place it. Choose level areas over sloping ones, and keep in mind that additional time and money will be required for any necessary grading.

2. Failing to Ventilate Properly

Insufficient air circulation can lead to your plants dying from disease pathogens that feed on moisture.

Even in the dead of winter, heat will rise inside a greenhouse without ventilation, possibly cooking your specimens on sunny days.

Vents are a great way to give your plants more airflow, whether they are located on the roof or in the side walls.

To get more ventilation, open them wider; to get less, close them. For as long as greenhouses have existed, this is a technique that has proven effective.

However, modern technology is your savior if you don't want to constantly run outside to open and close windows and vents.

These days, a lot of manufacturers provide automatic vent openers that take care of your vent adjustment!

They can run on electricity, which can come from solar power, AC power, or batteries. Alternatively, they can run on a wax that doesn't need a temperature sensor because it expands and contracts with heat.

If more ventilation is required, fans can be used. The fans create a vacuum by exhausting the hot air outside, which also pulls in cool air from the outside through open doors, vents, and any cracks.

The size, placement, and quantity of fans in a ventilation system are all determined by the ventilation requirements and size of your building; budget for whatever is required to prevent overheating.

3. Ignoring the pests:

    When working inside a structure, it's easy to forget about spooky things, but that would be a terrible mistake.

    Pests are drawn to greenhouses because of their warmth, humidity, and abundant flora, which all shout, "Come and get some!"

    Now that you've added vents that open frequently, they have easy access points. What, then, should a grower do?

    In general, you can avoid infestations in greenhouses in the same manner that you would in a garden.

    Regularly check your plants, make use of sterile instruments and growing medium, trim away any diseased, dead, or damaged plant tissue for disposal, and clear away any organic debris that accumulates.

    Other practices are unique to the greenhouse; these include maintaining functional screens and doors, removing standing water accumulation from floors, and maintaining weed-free, well-mowed grass surrounding the building.

    Physical, chemical, and biological controls are likewise largely comparable in both domains for dealing with existing pests.

    However, because you're working inside an enclosed area, you'll need to put on personal protective equipment (PPE) and take extra precautions to protect yourself from unnecessary sprays, as well as any nearby specimens, if needed.

4. Improper Fertilization:

Faulty fertilization is a common mistake made in the garden as well, and it can actually harm your plants or, at the very least, make your feeding ineffective.

Before you start, do some research and find out what your plants require to thrive.

Different kinds of species will have different kinds of needs. For instance, a nasturtium actually grows best in a lighter growing medium, but a calla lily requires a lot of extra nutrients to flower at its best.

Once you are aware of the dietary requirements of the plants you plan to grow, be sure to have a variety of fertilizers on hand for every specimen you end up cultivating!

You can utilize a marker system to identify each plant's needs, or you can group plants based on their requirements for fertilizer to keep everything organized.

A noticeable sticker, marker, or color-coded tag can serve as an instant reminder of the needs of your plant. Alternatively, record everything in your reliable garden notepad or journal.

5. Not providing the shade:

You should still protect your plants from the sun, even in the winter.

When direct sunlight hits a greenhouse, the heat it generates gets trapped inside, so it's helpful to have a method for shading your plants during particularly intense sunbursts so they can cool down.

Various shading materials can be purchased online and at do-it-yourself stores.

There are various fabrics with different weights, toughness, and prices, including polyester, PVC, polypropylene, and polyethylene.

The percentage of sunlight that is blocked by the degree of shading is typically expressed as a number between 10 and 90 percent.

You can install an electronic system that moves the shades automatically for you when a sensor detects a certain amount of UV exposure, or you can have them mounted on a pulley system against the inside of the ceiling. You can even pull them over the roof.

Naturally, an automated system will cost more than a manual one. But if you're searching for a convenient solution, automated shading definitely justifies the higher price.

6. Not Maintaining Your Greenhouse

You have to maintain the functionality of your greenhouse, or at the very least, prevent it from malfunctioning and becoming unusable.

As with your own home, it's a good idea to periodically check for possible structural problems.

Are there any areas that require lubrication, rusty spots, malfunctions, or cracks? These need to be resolved right away in order to avoid more serious issues later on.

Also, a lot of the greenhouse's components need to be checked.

A few examples include utility lines, power generators, irrigation hoses, tables, fan motors, shade pulleys, and the foundation itself.

Keeping the space clean is beneficial for safety, preventing pests and pathogens, and maintaining a clean appearance, in addition to repairing and preventing structural damage.

It might be surprising for you, but even a little effort in keeping the greenhouse clean will show you that it is contributing to the smooth functionality of the place.

7. Not Providing Heat

Night-time temperatures can drop significantly, particularly in the winter.

When the temperature drops below freezing, greenhouse heaters are required everywhere, even in places like South Texas and Florida.

Regular space heaters are not designed for use in hot and muggy conditions, so do not use them there. Instead, use one outside your home.

Use outdoor power cords and surge protectors that are rated for the outdoors, and only buy heaters that are specifically made for greenhouses.

8. Over- Or Under-Irrigating

No matter where you grow plants, this is a universal struggle. If you use too much water, you run the risk of suffocating the roots and encouraging the growth of diseases that prefer damp conditions.

If you give your specimens too little, they may wilt, wrinkle, and suffer potentially fatal cellular damage.

Irrigation can be made easier by grouping plants with similar water requirements together. You can also keep track of your plants' various water requirements by using the marking and logging techniques mentioned previously for fertilizers.

You have two options for watering your plants: either get an automated system that sprays, drips, or bottom-soaks them as needed, or use sprinklers and time to water them the old-fashioned way.

These systems can be useful for larger-scale operations because they make use of moisture sensors, timers, and/or pre-programmed settings.

9. Poor Plant Placement

The plants you have might be appropriate, but are they situated correctly? You might become aware of potential logistical problems after compiling your ideal list of plants.

Overly tall plants run the risk of colliding with the greenhouse ceiling and overshadowing nearby plants, which could result in less-than-ideal growth and the spread of disease.

Therefore, it's crucial to consider their mature sizes when placing them initially.

Plants with similar physiological requirements—such as those for sun, water, humidity, and fertilizer—should be grouped together.

Naturally, it will not be feasible to take into account every single factor, but attempting your best will help to ensure that each plant receives what it needs without too much trouble.

10. Too Much or Too Little Humidity

Just like with liquid water, an excessive amount of moisture in the air can let pathogens proliferate throughout your greenhouse.

However, the opposite issue could also arise: too little water could cause the plants to dry out quickly.

Plan ahead and combine plants that require similar levels of humidity.

Humidity can be effectively increased by misting plants with water and physically moving them closer together; it can be decreased by increasing ventilation, adding bottom heat, and moving plants farther apart.

When water is sprayed on the floor, it evaporates and increases humidity. This process, known as evaporative cooling, can be beneficial for overheated plants.

If your objective is to maintain a lower level of humidity instead, preventing excessive water spillage and creating a well-draining floor are both reliable ways to keep the ground dry.

11. Using the Wrong Growing Medium

Filling containers with regular garden soil could be disastrous in a greenhouse.

Compared to many commercial mediums, garden soil is less sterile and compacts more readily, which, over time, may kill the roots.

When compacted garden soil is placed in large containers, the top portion of the soil dries out, and the bottom of the container fills with water.

Garden soil containers may then become too heavy to lift or move safely.

Instead, use soilless growing media, which can avoid all of the aforementioned issues while also producing healthier plants and larger yields.

They are quite sterile, lightweight, and well-draining when used and formulated properly.

Depending on their unique requirements for moisture and nutrients, different plants have different requirements for media.

Nevertheless, you can at least easily make the required adjustments with a soilless media—amending garden soil is frequently far more difficult.

Conclusion

Now we know where to start to make a properly functioning greenhouse from its foundation till the stage when it is time to harvest your first crops.

The days when you felt de-motivated because of the fear of failing to make and maintain a reasonable greenhouse are gone. The wonderful thing about other people's mistakes is that we can learn from them without having to make the same ones twice.

It's acceptable if, however, you find that you are making a few mistakes of your own! I think it was well worth making those mistakes as long as you took something away from the experience.

If you've enjoyed reading this book, subscribe* to my mailing list to get exclusive content and sneak peeks at my future books.

Visit the link below:

http://eepurl.com/glvBjj

OR

Use the QR Code:

(*Must be 13 years or older to subscribe)